A LOOK AT THE HUMAN BODY

THE SKELETAL SYSTEM

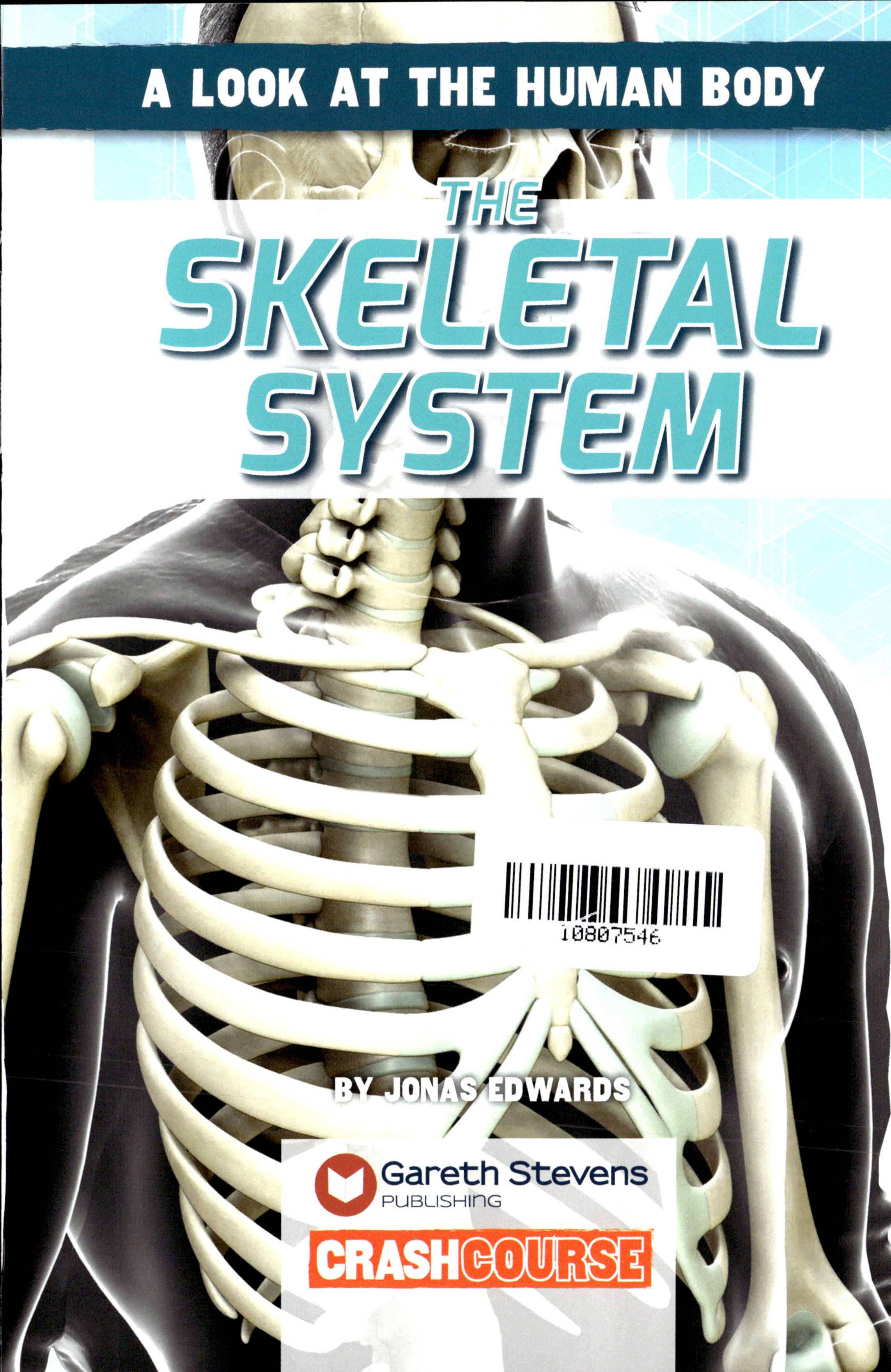

10807546

BY JONAS EDWARDS

Gareth Stevens PUBLISHING

CRASHCOURSE

Please visit our website, www.garethstevens.com. For a free color catalog of all our high-quality books, call toll free 1-800-542-2595 or fax 1-877-542-2596.

Library of Congress Cataloging-in-Publication Data

Names: Edwards, Jonas, author.
Title: The skeletal system / Jonas Edwards.
Description: New York : Gareth Stevens Publishing, [2022] | Series: A look at the human body | Includes index. | Audience: Grades 2–3 |
Identifiers: LCCN 2020030132 | ISBN 9781538266311 (library binding) | ISBN 9781538266298 (paperback) | ISBN 9781538266304 (6 pack) | ISBN 9781538266328 (ebook)
Subjects: LCSH: Human skeleton—Juvenile literature. | Bones—Juvenile literature. | Human physiology—Juvenile literature.
Classification: LCC QM101 .E39 2022 | DDC 612.7/51—dc23
LC record available at https://lccn.loc.gov/2020030132

First Edition

Published in 2022 by
Gareth Stevens Publishing
29 E. 21st Street
New York, NY 10010

Copyright © 2022 Gareth Stevens Publishing

Editor: Kate Mikoley
Designer: Andrea Davison-Bartolotta

Photo credits: Cover S K Chavan/Shutterstock.com; backgrounds Yuliia Petrenko/Shutterstock.com; p. 5 Lightspring/Shutterstock.com; p. 7 SciePro/Shutterstock.com; p. 9 Thongchai S/Shutterstock.com; p. 11 Samuel Borges Photography/Shutterstock.com; p. 13 (bottom) sirikorn thamniyom/Shutterstock.com; p. 13 (top) Krakenimages.com/Shutterstock.com; pp. 15 (left), 30 VectorMine/Shutterstock.com; p. 15 (right) sciencepics/Shutterstock.com; p. 17 Magic mine/Shutterstock.com; p. 19 studiovin/Shutterstock.com; p. 21 (bottom) itsjustme/Shutterstock.com; p. 21 (top) Jose Luis Calvo/Shutterstock.com; p. 23 (bottom) eranicle/Shutterstock.com; p. 23 (top) Sergey Novikov/Shutterstock.com; p. 25 Suzanne Tucker/Shutterstock.com; p. 27 Monkey Business Images/Shutterstock.com; p. 29 Jacek Chabraszewski/Shutterstock.com.

All rights reserved. No part of this book may be reproduced in any form without permission in writing from the publisher, except by a reviewer.

Printed in the United States of America

Some of the images in this book illustrate individuals who are models. The depictions do not imply actual situations or events.

CPSIA compliance information: Batch #CSGS22: For further information contact Gareth Stevens, New York, New York, at 1-800-542-2595.

Find us on

CONTENTS

Words in the glossary appear in **bold** type the first time they are used in the text.

LET'S GET SKELETAL!

What would your body look like without any bones? It would be a lot different, that's for sure! The collection of bones that makes up your body is called your skeleton. Without it, you'd just be a pile of skin and guts!

MAKE THE GRADE

The skeleton is what holds up your body and gives it a shape. It also covers important **organs**, keeping them safe, and helps you move.

Your skeletal system includes all the bones that make up your skeleton. But this system isn't just bones. It also includes body parts that support your bones and keep them connected, such as ligaments and tendons.

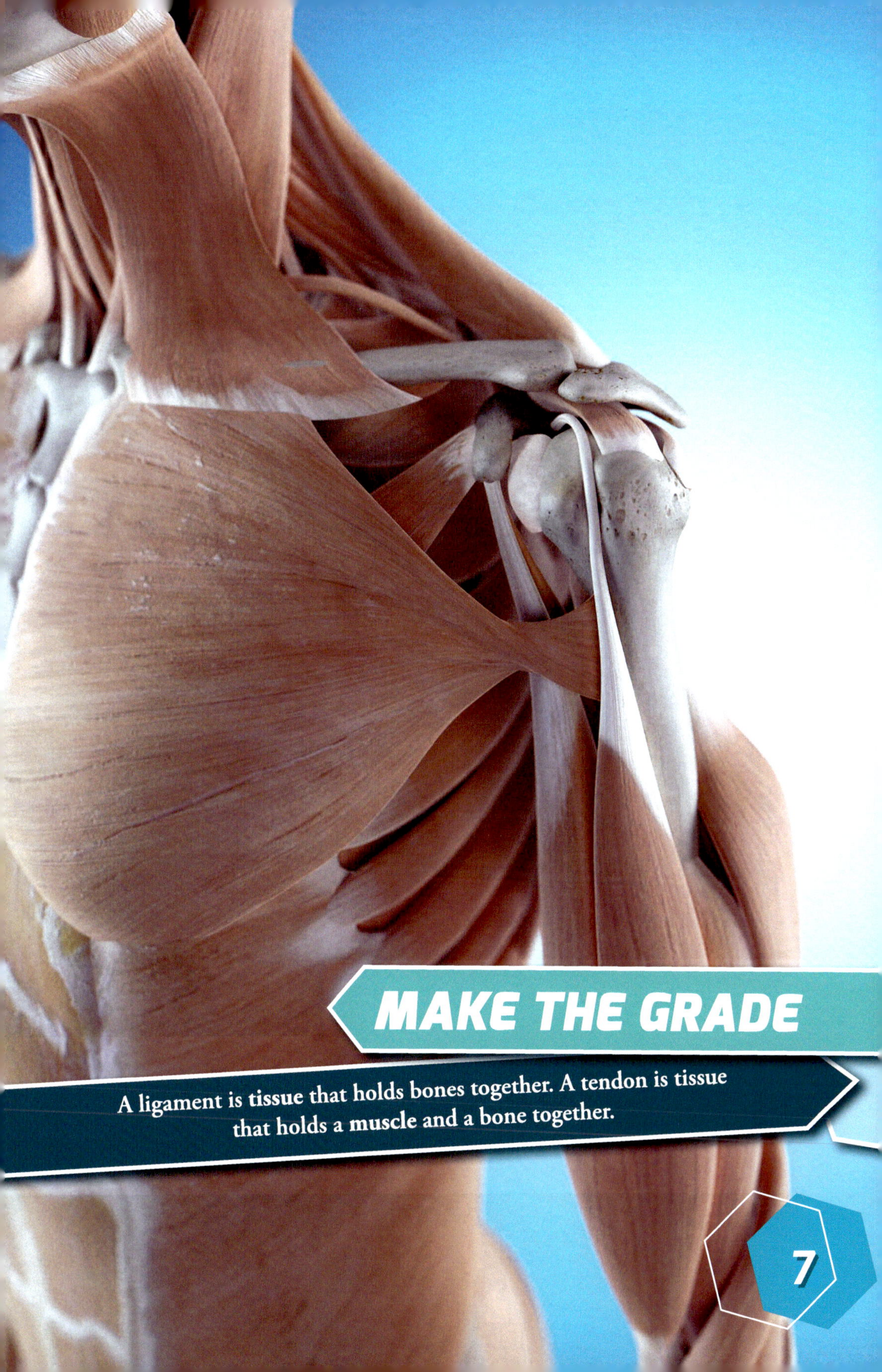

MAKE THE GRADE

A ligament is tissue that holds bones together. A tendon is tissue that holds a muscle and a bone together.

YOUR LIVING SKELETON

Skeletons in museums or classrooms can make bones seem lifeless. But your bones are very much alive—just like the rest of your body. They grow, work with other body systems, and can even fix themselves if broken!

MAKE THE GRADE

A doctor can give you a cast to help fix a broken bone, but it's really the bone that fixes itself. A cast keeps the broken bone from moving as new bone forms to fix the break.

YOUR GROWING BONES

When you were born, you had about 300 bones. As you've gotten older and bigger, this number has gone down. How is that possible? As you grow, some of your bones fuse, or grow together to form a bigger bone.

MAKE THE GRADE

By the time a person's bones stop growing, by about age 25, they're left with 206 bones.

In babies, some bones are made completely of cartilage. This is a kind of tissue that is soft and **flexible**, yet strong. Other bones are made partly of cartilage and partly of hard bone. As the bones grow, much of the cartilage becomes hard bone.

MAKE THE GRADE

Cartilage is part of an adult's skeletal system too. Body parts such as the ears and nose are made of cartilage. It also covers the ends of some bones, helping **joints** move smoothly.

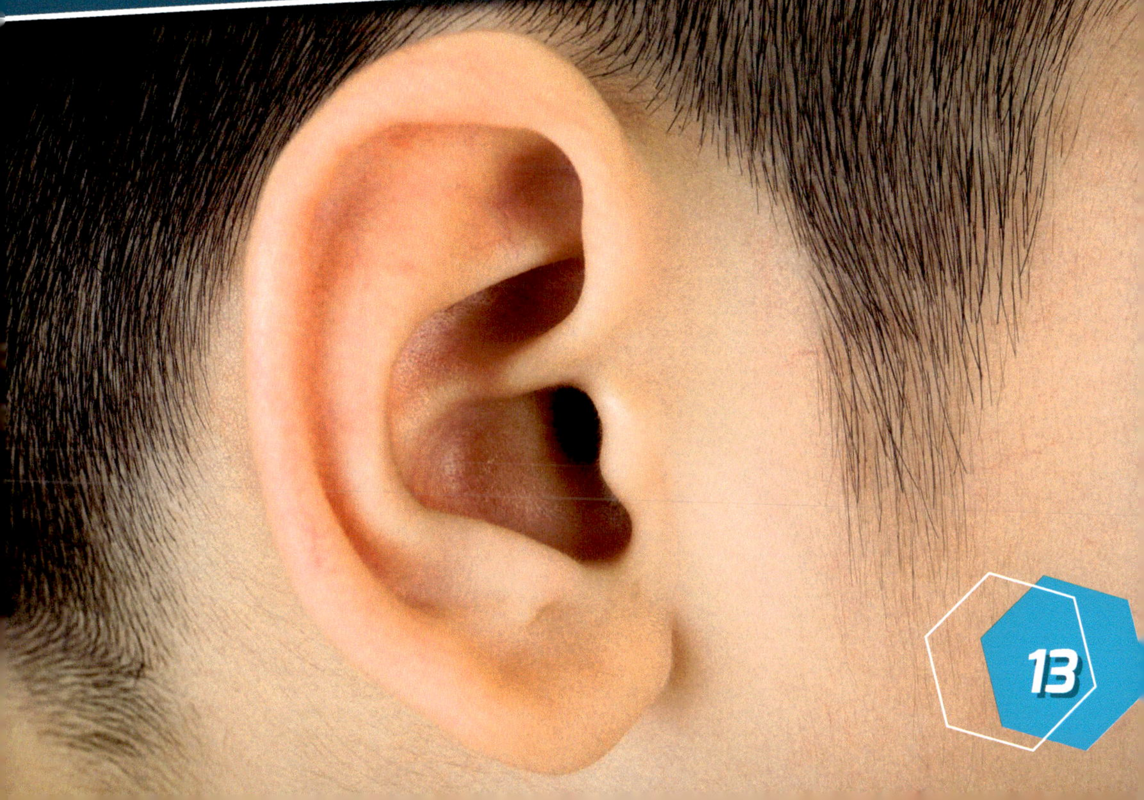

SEPARATING THE SKELETON

The skeleton can be grouped into two parts. The axial skeleton includes the bones that make up the head, called the skull. It also includes the spine and rib cage. These bones cover and **protect** the brain, spinal cord, heart, and **lungs**.

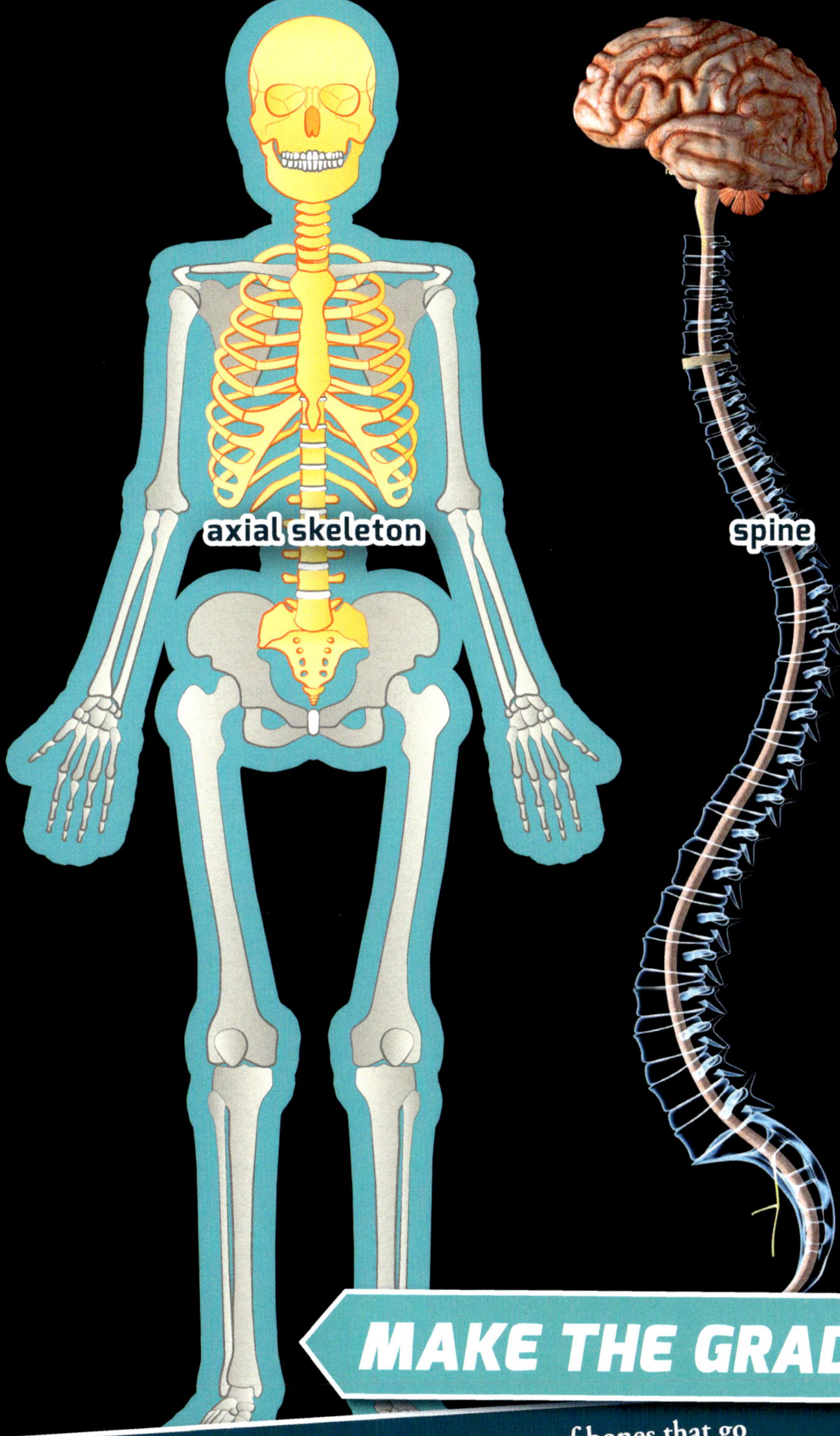

MAKE THE GRADE

The spine, also called the backbone, is a group of bones that go from the back of the neck down the back. They protect the spinal cord, a thick rope of nerve tissue extending from the brain.

The rest of the bones make up the appendicular skeleton. This includes the bones in your arms, legs, hips, and shoulders. These bones help you stand up, run, walk, and lift things up. Like the axial skeleton, they also protect organs.

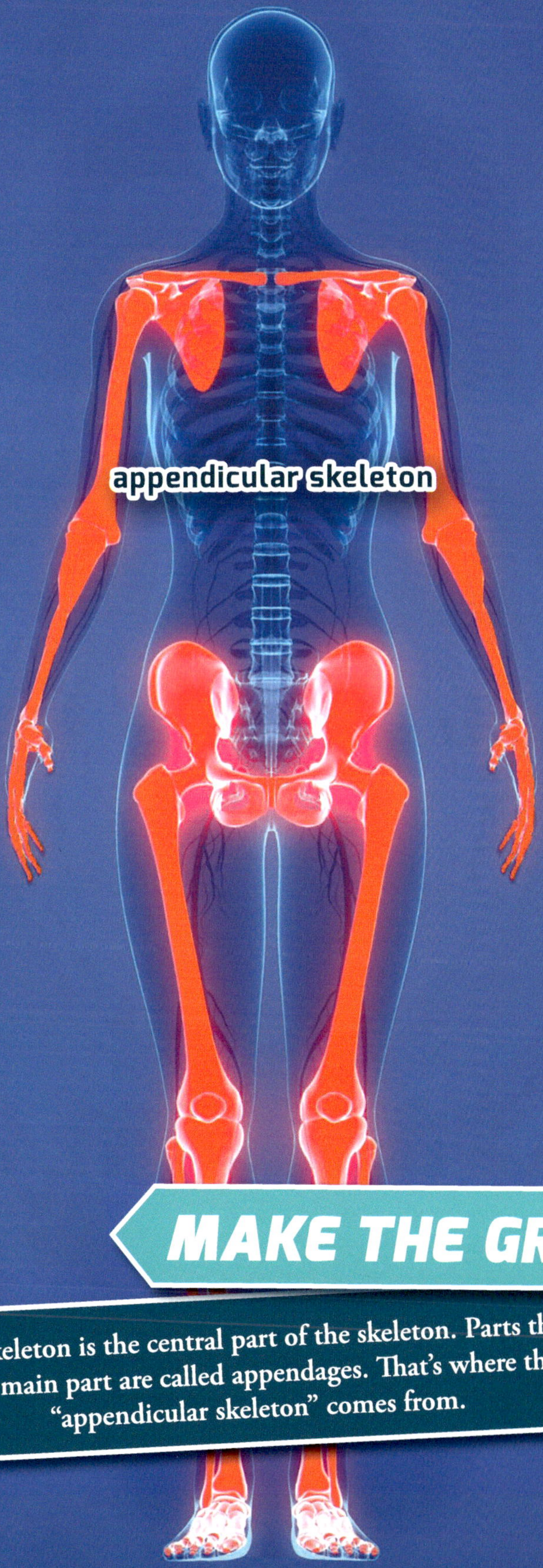

MAKE THE GRADE

The axial skeleton is the central part of the skeleton. Parts that come off of this main part are called appendages. That's where the name "appendicular skeleton" comes from.

BUILDING THE BONES

While different bones have different jobs, most of them are made of the same stuff. The outside part of the bone is a thin layer called the periosteum. Though it's thin, this layer holds nerves and small tubes that carry blood, called blood vessels.

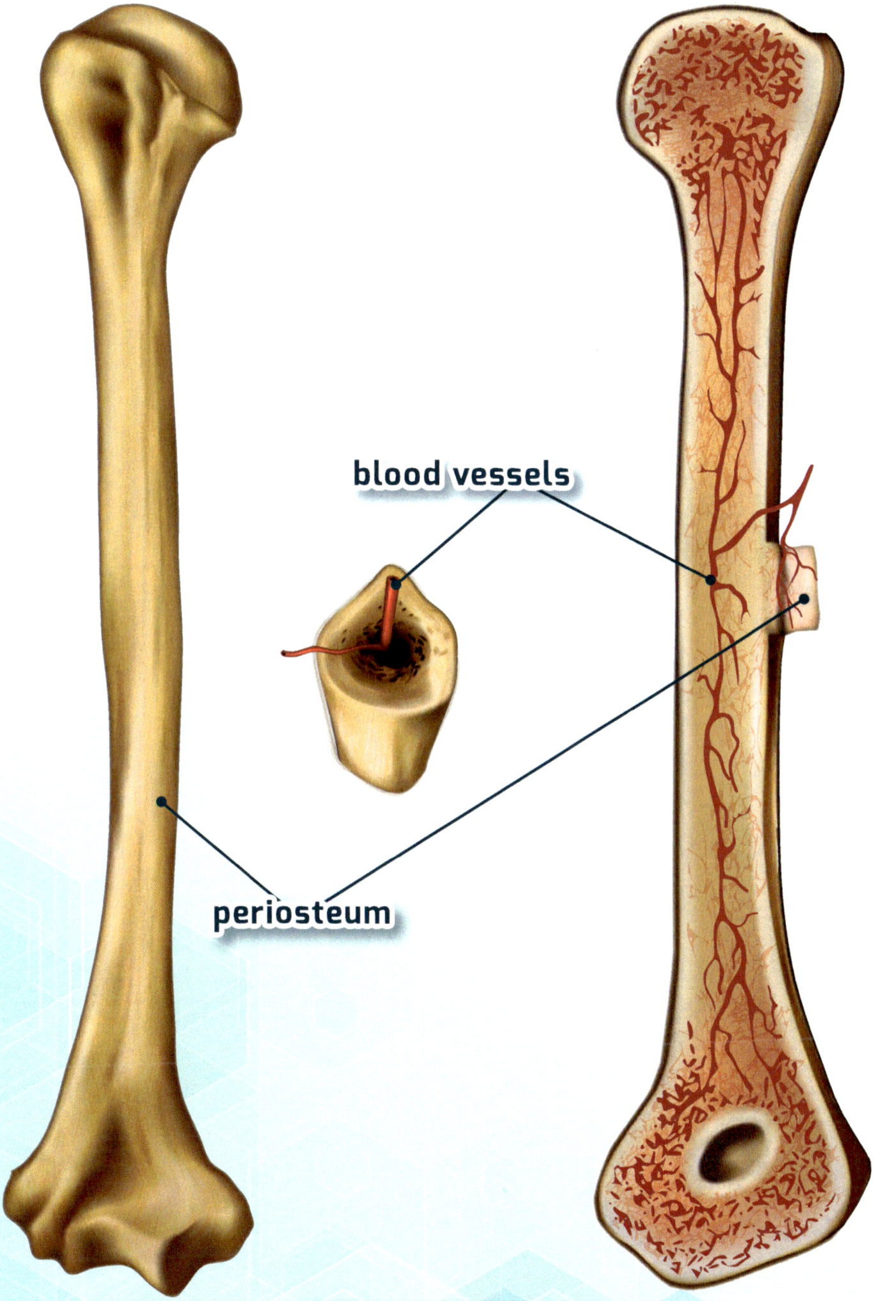

MAKE THE GRADE

Blood helps nourish the bones. The blood vessels in the periosteum bring blood to other parts of the bone too.

Beneath the periosteum is the compact bone. It is smooth and hard, and it makes up about 80 percent of each bone. Compact bone is strong, making it a good protector for the inside parts of the bones.

MAKE THE GRADE

Compact bone is mostly made up of cells called osteocytes.

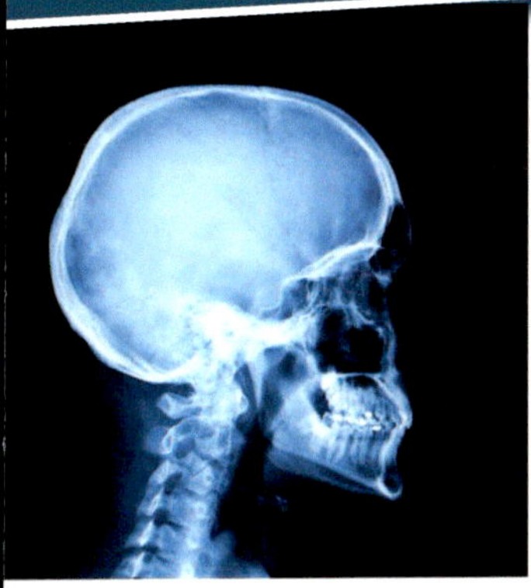

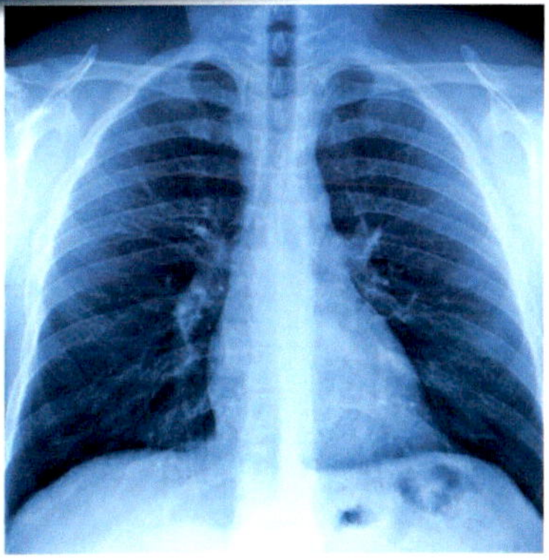

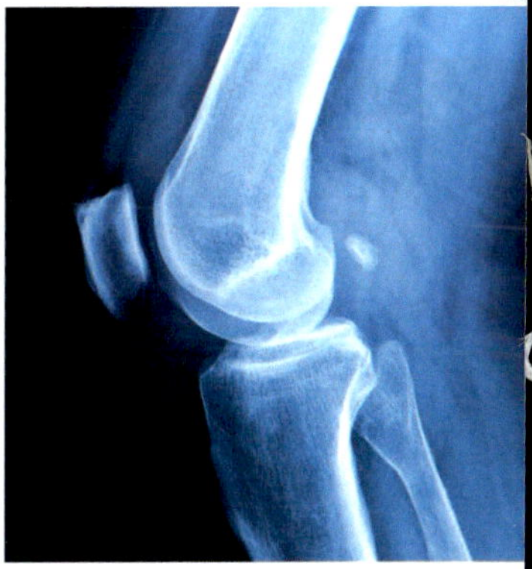

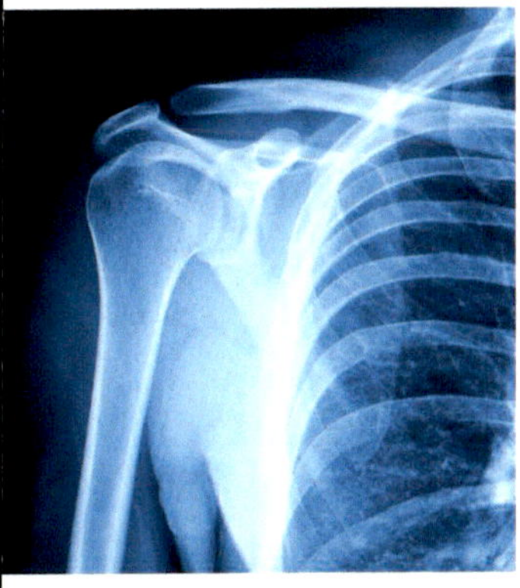

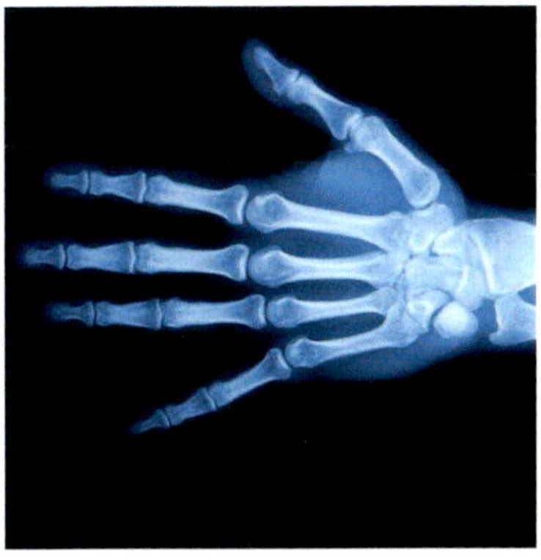

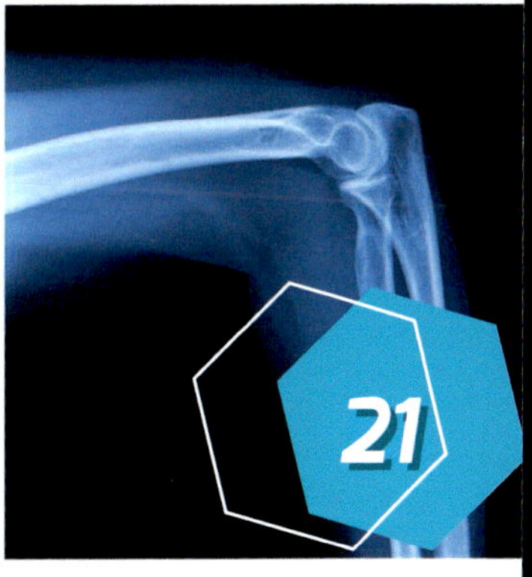

Under compact bone, you have spongy bone. This kind of bone has lots of little holes but is still very strong. It is filled with nerves and blood vessels. It also houses a soft matter called bone marrow.

MAKE THE GRADE

Spongy bone is often found at the ends of a bone. That's where most weight is put on the bone. Spongy bone helps bear this weight. This allows you to comfortably stand, run, or jump.

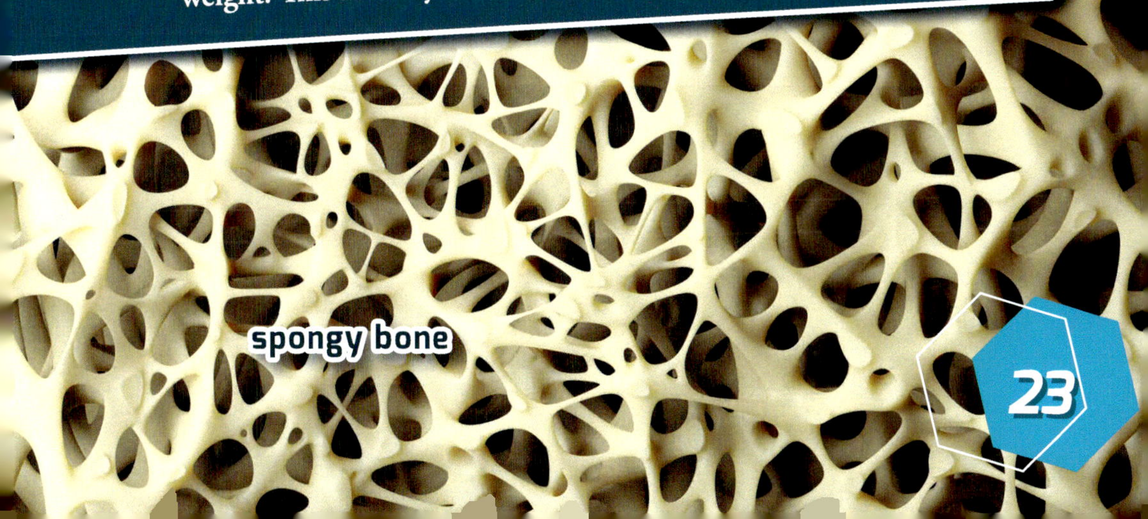

spongy bone

MARROW IN THE MIDDLE

Bone marrow feels sort of like jelly. You have red and yellow bone marrow, both found near the middle of bones. Red bone marrow helps your body make blood cells. Depending on the type, blood cells may fight **infection** or bring **oxygen** around the body.

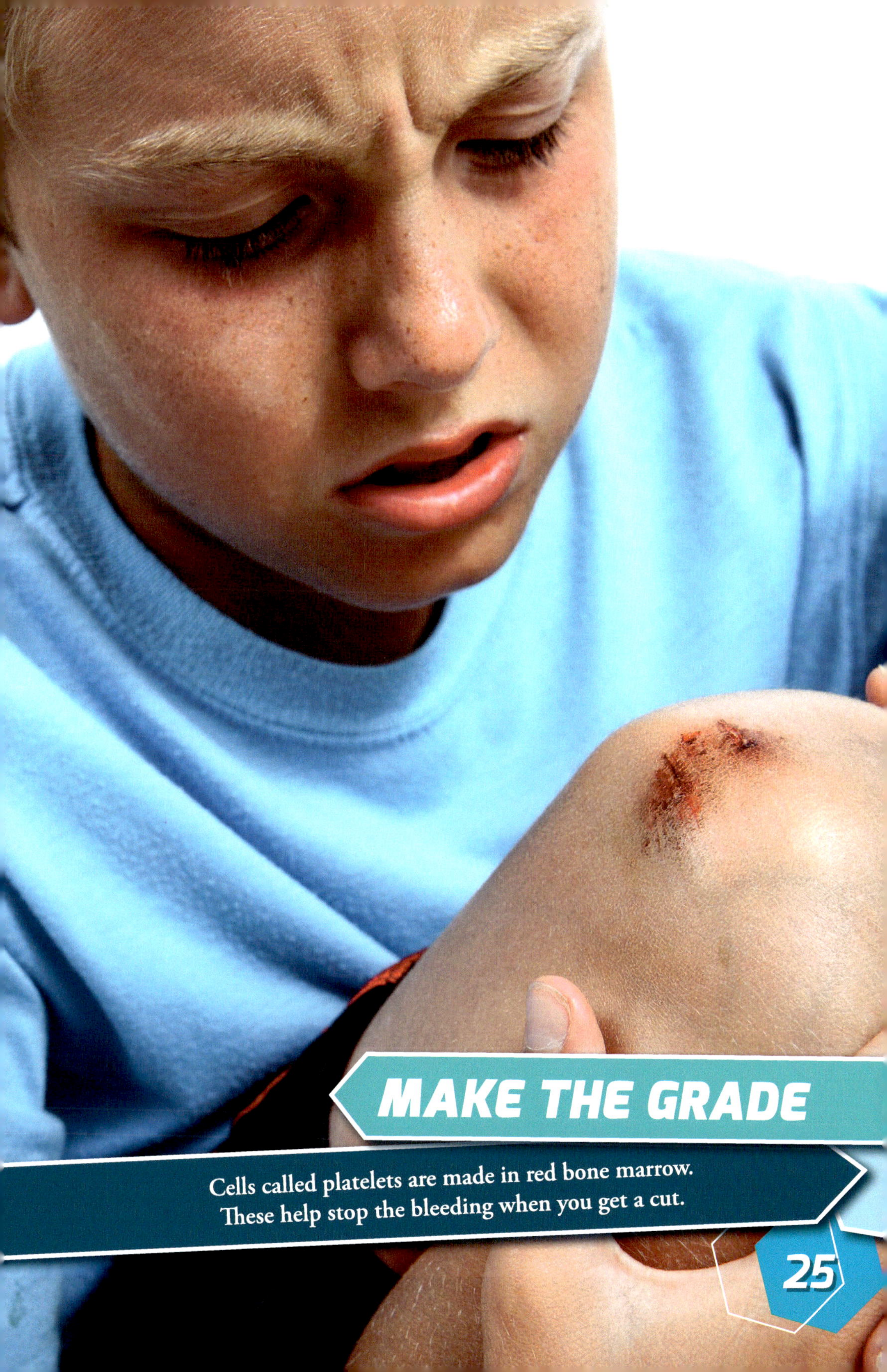

MAKE THE GRADE

Cells called platelets are made in red bone marrow. These help stop the bleeding when you get a cut.

Yellow bone marrow can be found in bones such as those that make up the spine, skull, arms, and legs. Its job is to store fat. As a person gets older, yellow bone marrow replaces most of their red bone marrow.

MAKE THE GRADE

Fat helps to give your body energy. Energy is the power to do things and be active.

SAFETY FOR THE SKELETON

Whether you're thinking about it or not, your skeletal system is always helping you. You need to help your skeletal system too. If you're riding a bike, skating, or playing certain sports, always wear a helmet and other gear to protect your bones!

MAKE THE GRADE

Being active, such as by playing sports, riding a bike, or dancing, can help make your bones stronger!

INSIDE A BONE

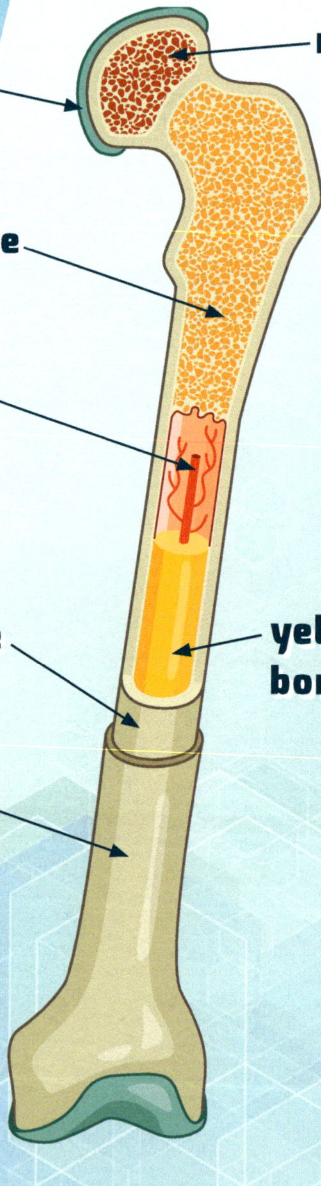

GLOSSARY

flexible: able to bend without breaking

infection: a sickness caused by germs

joint: a point in the body where two bones meet

lung: a part of an animal that takes in air when it breathes

muscle: one of the parts of the body that allow movement

nerve: a part of the body that sends messages to the brain and allows something to feel things

nourish: to provide with food or other things needed to live

organ: a part inside an animal's body

oxygen: a colorless, odorless gas that many animals, including people, need to breathe

protect: to keep safe

tissue: matter that forms the parts of living things

FOR MORE INFORMATION

BOOKS

Beevor, Lucy. *Understanding Our Skeleton*. Chicago, IL: Capstone, 2017.

Rustad, Martha E. H. *My Skeletal System: A 4D Book*. North Mankato, MN: Pebble, 2019.

WEBSITES

Bones, Muscles, and Joints
kidshealth.org/en/teens/bones-muscles-joints.html
Find out more about key parts of your skeletal system here.

Skeleton and Bones
www.dkfindout.com/us/human-body/skeleton-and-bones/
Head to this page to take a look at different parts of the human skeleton.

Publisher's note to educators and parents: Our editors have carefully reviewed these websites to ensure that they are suitable for students. Many websites change frequently, however, and we cannot guarantee that a site's future contents will continue to meet our high standards of quality and educational value. Be advised that students **should be closely supervised whenever they** access the internet.

INDEX